BREAD SCIENCE AND TECHNOLOGY

NIDHI GUPTA

(**B.Tech** *UDCT Mumbai, pursued MS, University of Maryland College Park, USA*)

INDIA • SINGAPORE • MALAYSIA

ISBN

Hardcase 979-8-89984-207-8
Paperback 979-8-89929-964-3

Table of Contents

Acknowledgment

This book is a dedication to my parents. I feel proud to have executed my idea to author a book and serves as a memento for them. Their constant support and motivation is what inspires me to write my first book as an author. I would also like to thank Dr. Jagdish Pai(ex-Exec Dir.- PFNDAI, ex-HOD-Food Engineering Department UDCT Mumbai) under whose able guidance I was able to not just author but also publish this book. Last but not the least my husband stands tall in always supporting me and serves as my pillar of strength.

CHAPTER 1

Wheat and its Characteristics

Introduction

Breadmaking requires a deep understanding of the complex phenomenon of process interactions and raw materials used. It is important to maintain consistency which characterizes the quality of bread products eg. loaf height and volume. Other parameters like smell and flavour are more difficult to define.

1.1 Wheat and its Structure

Wheat grain, is the principal cereal used for breadmaking. There are other cereals also used to some extent in breadmaking, but there are several properties in wheat which makes it suitable to be used for bread making. This chapter discusses the structural changes happening inside the bread while it is baked.

Structure of Wheat Grain

Average length of kernel is 8 mm. They are rounded on the germ side and have a longitudinal crease opposite germ which runs over entire length. It goes inside upto almost centre of grain and makes it difficult for miller to separate bran from endosperm.

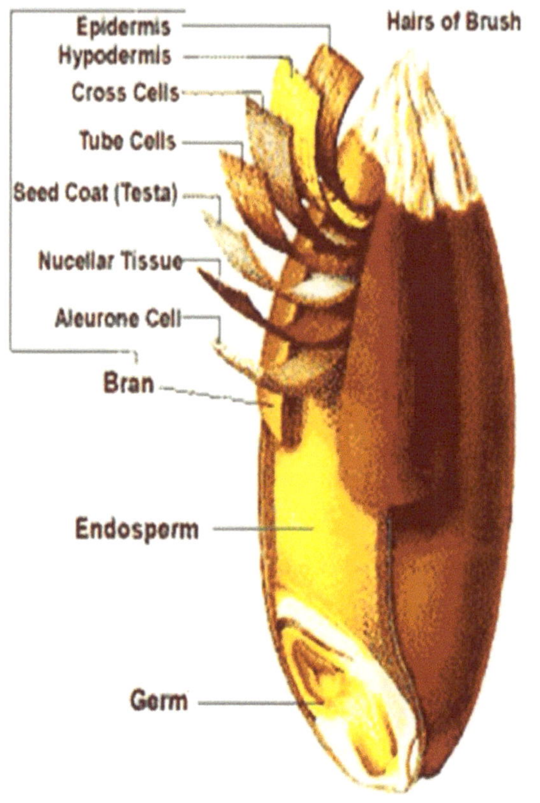

Fig. 1 Structure of wheat grain

A typical chemical composition of pericarp includes: - Proteins: 6%

Ash: 2%

Cellulose: 20%

Fat: 0.5%

Remaining are NSPs (Non- Starch Polysaccharides)

1.2 Parts of Wheat Kernel

1. **ALEURONE LAYER:** This is a one cell thick layer at the inner side of the bran. It is rich in minerals, vitamins, and phenolic compounds and lignins of the wheat kernel. The fibre rich aleurone layer can be separated from other layers of wheat bran

and helping in obtaining a fibre rich concentrate with whole grain kernel bioactives. Novel milling technologies and dry fractionation technique have made it possible for full separation of aleurone cells from other layers of wheat bran. The role of dietary fibre in health and disease have triggered interest in separation of the micronutrient rich aleurone fraction from wheat.

2. **GERM (EMBRYO):** Consists of 2 parts – Embroyonic axis, Scutellum Germ is high in protein (25%), sugar (sucrose and raffinose) (18%), oil (16% of embryonic axis) and 32% of scuttelum are oil and ash (5%). It lacks in starch and is high in B vitamins and many enzymes. Germ is also high in vitamin E content (500 ppm). As Germ contains oil, it reduces the keeping quality of flour hence it is desirable to remove germ during the milling process.

3. **ENDOSPERM:** Starchy endosperm is made up of 3 types of cell

   Outer (peripheral cells) Middle (Prismatic cells)

   Inner (Central cells)

   The cell walls of endosperm are composed of pentosans, other hemicelluloses and Beta-glucans but not cellulose. Thickness of cell walls varies as they are thicker near aleurone layer and vary between soft and hardwheats.

   The endosperm is the source of white flour. It contains the greatest share of proteins, carbohydrates and iron, as well as major B vitamins, riboflavin, niacin and thiamine. It is also a source of soluble fibre. Endosperm makes up around 83% of a kernel of wheat.

1.3 Wheat Types

Hard Wheat: High water absorption due to thick cell walls hemicellulose absorbs large amount of water.

Soft Wheat: Low water absorption due to thin cell walls.

Proteins and Starch are tightly adhered to each other and protein appears to coat the surface of starch. Hence hard wheats fracture along cell walls rather than through cell contents when milled and break through starch granules rather than between starch protein interfaces.

The endosperm is the biggest part of the kernel and is the source of refined wheat flour. The aleurone layer surrounds the endosperm tissue of wheat seeds. The germ is the nutrient rich embryo which will grow into a new wheat plant.

Hard Wheat

Hard wheat (T.aestivum) with strong and extensible gluten and high protein is required for making good bread. For pasta products, hard wheat (T.durum) with strong gluten, high protein, low yellow berry incidence and high yellow pigment (β-carotene) content are required.

In the US, both hard and soft wheats are hexaploid while durum is tetraploid. In Europe, durum(tetraploid) is considered to be hard and all other wheats (hexaploid) are soft. Thus, Europe considers US hard wheats to be soft wheats.

Soft Wheat

Soft and very soft varieties of wheat offer lower protein contents (8-12%) and have weak gluten network. The softer variety of wheat produces flour suitable for biscuits, cakes, white noodles, pastries, cookies etc. Two types of soft wheat are grown in the United States: soft red winter and soft white. Soft red winter wheat is grown where higher rainfall and humidity prevail. Soft white wheat is winter sown and have a "vernalization" requirement, which must be fulfilled to complete their reproductive cycle.

Promising varieties of wheat products

Based on analysis of AICW&BIP trials, product specific varieties are identified for chapati(>8.0 score out of 10.0), bread(>575 mL loaf volume), biscuit(>11.0 spread factor) and pasta(>7.0 score out of 9.0). Chapati quality in India is among the best in the world but improvement in bread and biscuit quality in needed.

1.4 Some Common Milling Product Terms Used by Bakers

Bleached Flour	Flour is chemically treated to improve baking quality and colour
Composite Flour	Flour made by blending varying amounts of non wheat flour and wheat flour and used for the production of baked goods.
Farina	A very pure wheat endosperm about the granulation of medium sizing, endosperm particles from hard wheat can be used for breakfast cereal.
1st clear flour	Portion of flour remaining after a "patent" cut of flour has been taken off, clear flour is high in ash content and protein than patent and marketwise is secondary in value.
2nd clear flour	Lower grade portion or division of clear flour from the tail end reductions of the milling system having a higher ash and poorer colour than those included in first clear.
Germ	The embryo extracted from the grain kernels.
Hard Flour	Flour is strong in character requiring prolonged fermentation to ripen and has considerable fermentation tolerance and stability.

Patent Flour	The most highly refined flour (combination of four streams) from the front of the mill, lower in ash and protein with good dress and colour and considered highest in value. SPF (Short Patent Flour) is more highly refined than long patent flour (LPF).
Soft Flour	Flour with a low percentage of weak gluten, milled from soft red winter or soft white wheat.
Treated Flour	Flour to which supplement has been added such as vitamins, calcium, iron, self rising ingredients etc.

1.5 Science Behind the Use of Wheat in Breadmaking

Wheat is comprised of endosperm (85%), Bran (12.5%) and Germ (2.5%).

Wheat flour contains gluten proteins which is required for the structure of baked goods. Glutenin and Gliadin are the main proteins involved in gluten network formation in baked goods.

For bread making, wheat flour, water, yeast and salt are combined. Other ingredients can also be added like other cereals, fat, milk and milk products, emulsifiers, fruits, gluten to add value to the bread. There are majorly three processes happening on combining these ingredients:

1. The protein in the flour starts to hydrate i.e. combines with some water to form gluten network. Flour consists of discrete particles while gluten is cohesive forming a three dimensional structure which binds the flour particles together in dough. Gluten is extensible in nature and possesses a degree of recoil.

2. Air bubbles get trapped inside the gluten network as the dough is handled and the bubbles divide or coalesce.

3. Enzymes in the yeast start to ferment the sugars present in the flour and sugars released by diastatic action of the amylases on damaged starch in the flour breaking them down to alcohol and carbon dioxide. The carbon dioxide gas mixes with bubbles and brings an expansion in the dough.

Due to the formation of gluten network , wheat is the preferred cereal in bread making.

Wheat with high protein content are typically:

Grown in hot climates, have a glassy and vitreous look, are classified as hard wheat and are better for breadmaking.

1.6 Wheat Flour Composition

Wheat Flour is composed of Glutenin and Gliadin proteins. Glutenin provides elasticity and strength. Strength is the force required to stretch the dough. Elasticity means the ability to bounce back once flour is stretched. Gluten is a water insoluble protein which is formed when water is mixed with wheat flour. When sufficient water is added to dry flour, the two protein glutenin and gliadin, emerge from a "frozen-state" and become flexible. As water and flour are mixed, the hydrated proteins are brought together and begin to interact.

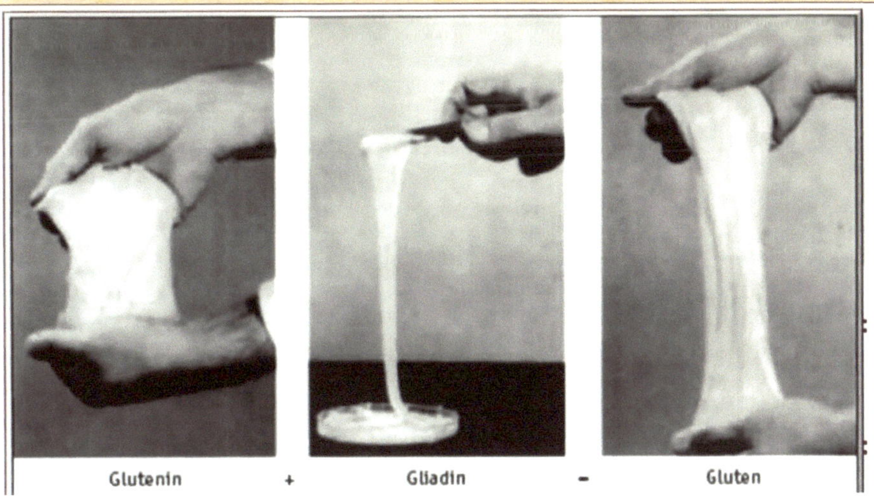

They begin to stick with each other with the formation of chemical bonds. These new chemical bonds are called cross-links. In case of gluten, a number of different chemical bonds form between proteins, with some disulfide bonds being stronger than others.

The mechanical shear causes the gluten bonds to form and become a viscoelastic matrix holding starch granules in flour. Hence, only a batter or dough can contain gluten, not raw flour.

1.7 Quality Tests for Flour

Chemical Analysis: Moisture, Ash, Protein, Sedimentation Value, Gluten, Alcoholic Acidity, Acid Insoluble Ash, Falling Number.

Water Absorption Power – WAP: is the quantity of water required to get dough of desired consistency, expressed as percentage of flour weight. Higher WAP meas better bread quality and increased product yield. Desired WAP- 67 to 72% in white bread production.

Dough Rheological Properties are also studied using farinograph, extensograph and alveograph.

Farinograph is a recording Dough mixer. It measures and records the resistance offered by dough against mixing blades operating at constant speed and temperature.

An **Extensograph** can be used to stretch dough until it breaks and study the extensibility and resistance to extension.

Area under the curve (cm^2)	Dough Strength
80	Weak
80-120	Medium strong
120-200	Strong
>200	Extra strong

Area under the extensograph curve with dough ***strength***

Alveograph is a rheological tool used to assess baking performance of flours used in baked products (bread, noodles, tortillas, biscuits etc.) Alveograph data is used to gain understanding of fluctuations in dough rheological changes by assessing tenacity, elasticity, baking strength, resistance of dough to deformation, extensibility.

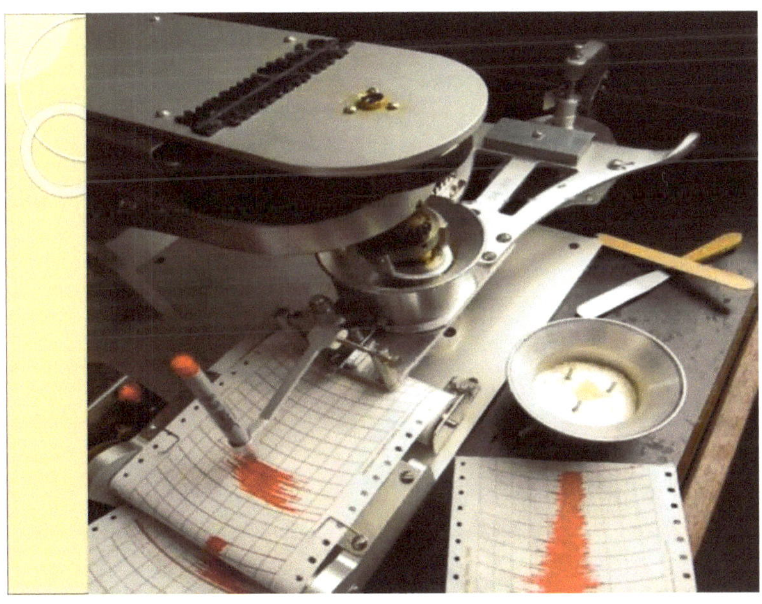

A pictorial representation of a Farinograph

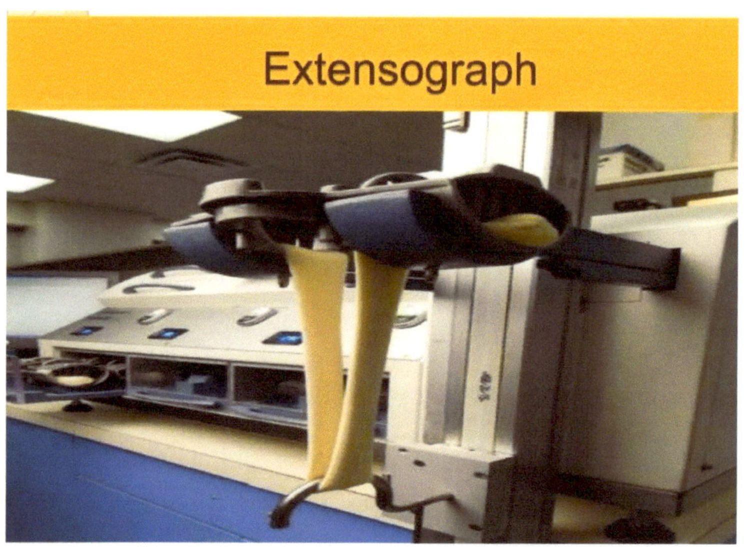

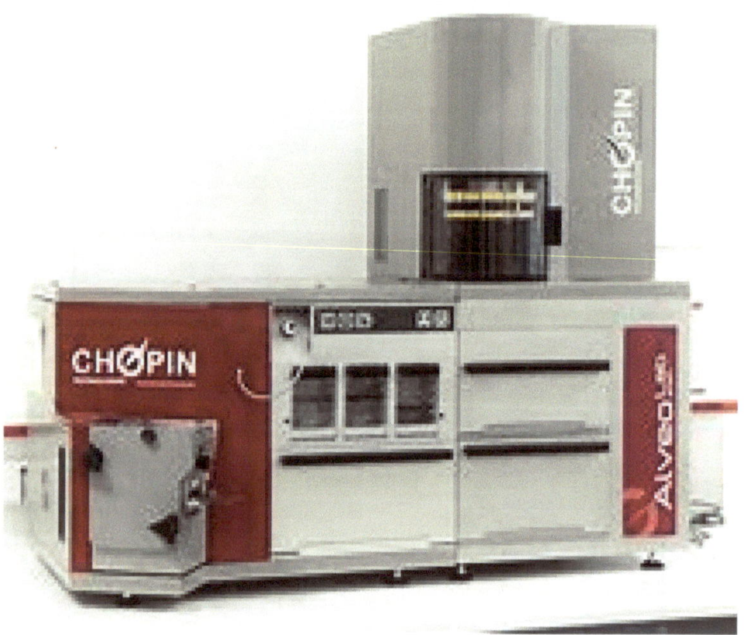

Choplin Alveograph

CHAPTER 2

Bread Processing and Breadmaking Methods

2.1 Bread Quality Parameters

FLOUR

A good quality flour should have:

- Adequate amount of protein which yields gluten on hydration having satisfactory elastic properties.
- Levels of amylase and damaged starch are adequate to yield sufficient sugars for yeast fermentation

Satisfactory moisture content not higher than 14% to permit safe storage.

YEAST

The quantity of yeast used is inversely proportional to the duration of fermentation. Yeast works as a leavening agent by providing CO_2 , it also affects the rheological properties of dough through lowering of pH by CO_2 production, evolution of alcohol and mechanical effects of bubble expansion. Yeast has a significant effect on aroma and flavour of bread.

AMYLASE

Normal flour contains a good quantity of beta-amylase and lower amounts of alpha-amylase. However, alpha-amylase amount exceeds when wheat germinates. Hence, flour made from sprouted wheat grains contains too high alpha-amylase with the result that during baking, some starch is changed into dextrin resulting in reduced water holding capacity. Dextrins make crumb sticky.

FAT

Fat is an important ingredient in breadmaking as it improves loaf volume, reduces crust toughness and gives thinner crumb cell walls resulting in soft textured loaf. Fat keeps bread soft and palatable for longer time.

SUGAR

Sugar is added to bread giving an acceptable sweet flavour.

Vital **Gluten**

Vital Gluten is used in a manner where it retains its properties to absorb water. It is used to fortify weak flours by improving the protein content of the flour, in starch reduced high protein breads in which gluten acts as a source of protein and texturizing agent, in high fibre breads to maintain texture and volume.

2.2 Colour of Bread Crust and Crumb

The brown colour in bread is due to the production of melanoidins formed by non- enzymatic browning reaction (Maillard reaction) between amino acids, dextrins and reducing carbohydrates. Addition of amino acids to flours gives pale crust colour resulting in improvement of colour.

2.3 Aroma of Bread

The aroma of bread is due to the interaction between reducing sugars and amino acids accompanied by formation of aldehydes. The flavour of bread is due to the crust mainly. During the natural fermentation process, which occurs in breadmaking new flavour products are generated. Both the intensity of these flavours and developed 'flavour' notes change with increased fermentation time. Flavour changes associated with the development of acid flavours from microbial activity in the dough are readily detected in the flavour of the bread crumb. Several hours of fermentation are required before any significant changes in the flavour profile of bread crumb. When breadmaking process has no provision for lengthy fermentation times, it is often the practice to develop flavour in a 'pre-ferment', 'brew' or 'sponge' which is mixed with remaining ingredients to form dough for final processing.

During baking, old flavour compounds are lost and many new flavour compounds are formed. Formation of a dark, mostly brown crust on the outer surface of the dough takes place. These changes are associated with complex processes commonly known as 'Maillard Browning.' Several views have been expressed that 80% of the bread flavour is derived from product crust.

Bread flavour is strongly influenced by ratio of crust to crumb. Development of Maillard browning products during crust formation influences the product flavour, UK sandwich read carry different flavour profile compared to French baguette. In case of baguette, proportion of crust to crumb is much higher so that a larger quantity of compounds contribute to product flavour. Lower proportion of flavour compounds in UK sandwich bread are seen as being detriment but the character of bread is aimed at completely end use to that of baguette.

2.4 Fermentation

Starch in the flour is broken down to disaccharide maltose by amylase enzymes, the maltose is split into glucose by maltase, glucose and fructose are fermented to CO_2 and alcohol by zymase complex.

(Zymase is the name given to a group of fourteen enzymes used formerly)

While milling, some starch granules are broken down and these granules are termed as damaged starch. It is these damaged starch which are attached by amylases to yield sugar needed during fermentation. When the amylase enzyme breaks down the damaged starch, water bound to starch is released which causes softening of the dough. Excessive level of damaged starch can adversely affect the bread quality by decreasing the loaf volume and making the bread look less attractive in appearance.

2.5 Structure of Bread

How are breads different from other bakery products like cakes, pastries, biscuits etc.? Let us try and understand the same under in this discussion.

As we know by now, bread requires the formation of gluten to trap the gas generated by the addition of yeast. Gluten is formed on mixing of wheat flour and water. Salt is also added to give more flavour to the product. However, bread develops flavours in the crust during baking as wheat flour and water are mixed with or without yeast.

Ingredients added to bread also confer bread its definition and characteristics. Cakes, pastries and some biscuits contain sugar which is absent in bread. However, there are sweet breads available in USA which contain sugar and also sweet bread market exists in India.

The other ingredient which differentiates bread from cakes, pastries and biscuits is fat. Most breads contain much lower proportions of fat as compared to cakes, pastries and biscuits.

A distinctive characteristic of bread is its gluten network. Gluten formation does not occur in cakes and is actively discouraged by addition of sugar and fat. Most biscuits and pastries also have limited gluten formation on comparison with most breads. Addition of right amount of water is also essential for the formation of gluten and for modifying the dough rheology. Too much or too little water cannot form the optimum gluten qualities in order to trap the gases from yeast fermentation.

Hence bread characteristic is different from other fermented products. The formation of a gluten network in the dough helps in the formation of a cellular crumb structure which attributes to texture and other eating qualities different from other baked products. On pressing this crumb structure and releasing the pressure, it springs back to assume its original shape. This combination of a cellular crumb with the ability to recover after being compressed, distinguishes breads from other baked products and these very characteristics bakers try to achieve in most breads.

2.6 Dough Mixers

Dough Mixers fall in four categories:

- Horizontal Mixers
- High Intensity Mixers
- Spiral Mixers
- Vertical Mixers

Horizontal Mixers:- A horizontal mixer gives a clear advantage of mixing thousands of pounds of dough in a short period of time. Medium to large size bakeries depend on this type of machine due to its sheer size and capability. This type of mixer is perfect for mixing things like short doughs, tortillas, fillings, pie doughs, health breads, cheesecake, muffins. These mixers come with mixing arms which allows to customize based on type of dough.

Spiral Mixers:- *Spiral Mixers are* for mixing varieties of dough. They come with two motors, one to turn the mixing tool and one to turn the bowl, mixing dough more efficiently and more gently than a horizontal mixer. Spiral mixers use an energy efficient mixing process, making the cost to run them lower than that of horizontal mixer.

However, Spiral mixers are only meant for bread doughs.

Accessories and **attachments**

Dough **Hook:-**

- Used for mixing most bread, roll and pizza dough which requires folding and stretching action.

- These agitators are useful for use on all yeast raised doughs.

Flat **Beater:-**

Provides a wide, flat surface which is perfect for mixing batter for cakes and brownies and also firm to hard cookie dough.

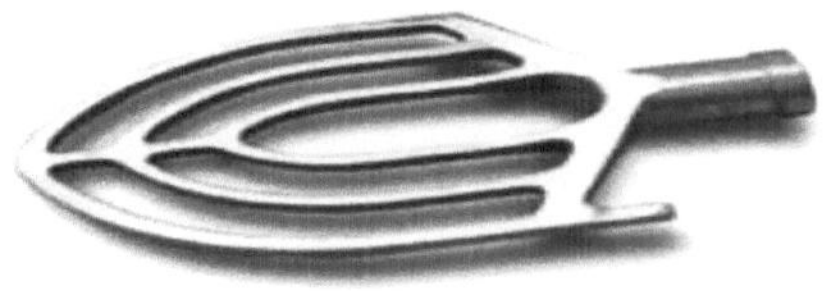

Wire **Whip:-**

Wire whip is made of wire strands arranged in an oval shape. It is designed for maximum blending of air into light products like custards, puddings, and whipped cream.

Bowl **Scapper:-**

It is used to continuously scrape the sides of the mixing bowl as the agitator turns.

This is a great way to minimize ingredient waste and reduce bowl scraping

Factors affecting Dough Fermentation:-

- Amount and type of yeast
- Dough Temperature
- Room Temperature
- Humidity

- Fermentation Time
- Amount of Salt
- Amount and Type of Sugar
- pH of dough
- Consistency of dough

Some of these factors are discussed below:-

Amount and type of yeast:- The first fermented bread doughs were inoculated with various wild yeasts and by saving small portions of the raw dough and feeding it more flour and water, the bakers of antiquity were able to create leavened breads on a regular basis. Most of the yeast cells in a sourdough starter originated from the outside of the bran layer. By 1920s, bread made with commercial yeast- Saccharomyces Cerevisiae was sold alongside traditional loaves made from wild sour cultures. Scientists isolated and successfully cultivated certain strains of saccharomyces cerevisiae which could act faster than others, thereby reducing the time necessary for dough to buildup sufficient gas before shaping.

Water:- Just adding or subtracting water from a dough or preferment can have profound effects on the rate and type of fermentation. The structure of bread dough can be manipulated by changing its water content. An excess of water creates an ideal environment for enzyme activity which weakens the gluten bonds and creates stickier dough. Dry dough tend to be stronger with a close cottony crumb while extra water can ensure an extensible dough with large irregular holes.

The mineral content of water can give dough somewhat greater strength, and those same minerals are actually beneficial to fermentation. A lack of minerals in soft water tends to lengthen fermentation times and may give dough an unusual level of extensibility.

A lack of strength from too soft water in bread dough can be addressed by extending fermentation times or adding sets of fold during bulk fermentation.

Salt:- Salt can certainly be used to enhance flavour of other dough ingredients but its effect on various dough processes are numerous and should not be overlooked.

It ***tightens the gluten structure*** and adds strength. This improves the dough ability to capture and hold carbon dioxide gas from fermentation, and ensures loaf will have good volume. Salt is a natural antioxidant, and if we subtract salt from a bread dough being mixed at higher speed, the rate of oxidation in the dough is dramatically increased. Most of the carotenoid pigments may be destroyed before salt is added, thereby significantly reducing the flavor and aroma components associated with them.

Even small amount of salt slows rate of fermentation and enzyme activity in bread dough and pre-ferments. This is because salt crystals are *hygroscopic* which means they tend to draw water away from their environment. When salt crystals and yeast cells are competing for same water, salt wins, which slows yeast life processes significantly.

Addition of small amounts of salt is useful in avoiding overfermentation in preferments but the final level of salt in bread dough is typically limited to 1.8-2.2% of flour weight with 2% being the norm.

Fermentation Time:- Yeast cells produce more sugars than their bacterial counterparts so their numbers within dough grow more quickly. Wild yeast in sour cultures work more slowly than their commercially made cousins but even they can surpass bacteria in a race to consume sugar. A bread dough made with manufactured compressed yeast at about 1% of weight of the flour used needs only 1 ½ to 2 hours to produce sufficient carbon dioxide to leaven the dough. With lactic acid bacteria, the dough needs 3 to 4 hours to accumulate to produce enough acid to produce bread dough taste very good.

Fermentation Temperature:- Increasing the temperature of bread dough or of the environment in which it is fermented has a dramatic effect on yeast fermentation by increasing the rate at which yeast cells consume simple sugars. The effect of increased temperature on bacterial fermentation is to increase the proportion of homofermentative bacteria and the by-product they produce which is the milder lactic acid. So increasing the temperature of dough or its environment speeds alcohol and gas production while maintaining a milder flavor profile.

2.7 Breadmaking Processes

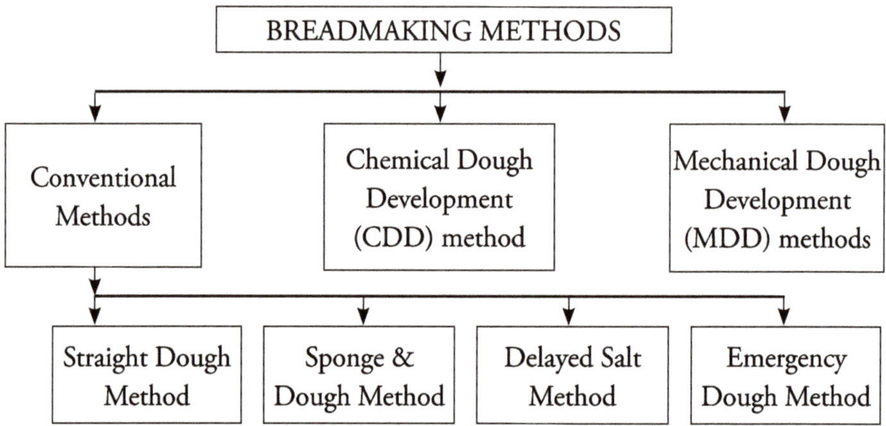

1. Straight Dough System:-

All ingredients (dry and liquid) are placed together in mixer and dough is then mixed to full development. Unlike a bulk fermentation in the Sponge and Dough system, this process does not use any fermentation step post mixing. The goal of straight dough system is obtaining high quality and standardized bread in very short time (3-4 hours from scaling through packaging compared to 6-8 hours in the sponge and dough system). Bakers are thus able to comply with unexpected customer orders and offer better service.

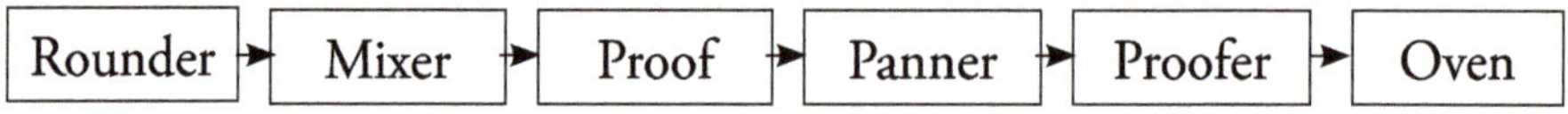

Yeast is used for leavening and dough is acidified by addition of lactic acid or acidic citrate. Dough is mixed slowly to prevent toughening which could be used by high viscosity of the pentosans. The lowering of pH has a marked effect upon the hydration and swelling of gluten, rate of enzyme action and other chemical reactions involving organic compounds.

Lowering of pH and action of proteolytic enzymes on protein, production of organic alcohols etc., alters the colloidal structure of proteins in dough. This helps to increase the extensibility and elasticity of gluten, thereby enabling to form thin gas retaining cells which can expand without rupturing. A dough with maximum gas retaining capacity and has developed maximum elasticity and springiness is said to be mature.

2. Sour Dough Process:-

Sour Dough is a sponge and dough process. Starter dough is prepared by allowing a rye dough to stand at 24^0-27^0C for several hours to induce a natural lactic acid fermentation caused by grain microorganisms. Alternatively, the rye dough is inoculated with sour milk and rested for few hours after which a pure culture of organic acids (acetic, lactic, tartaric, citric, fumaric) is added to simulate flavour of normally soured dough.

Flavour of San Francisco sour dough bread is largely due to lactic and acetic acids produced from D-glucose by *Lactobacillus,* a bacterium active in sour dough starter.

All doughs containing rye flour require acidification because sour conditions improve the swelling power of pentosans and partly inactivate amylase, which would have a detrimental effect on the baking process and impair normal crumb formation. There are certain advantages of using this method like less need of yeast as it multiplies during sponge fermentation, increased flavour of bread as it is developed by long fermentation process.

3. No Time Dough Method:-

Even though the dough is fermented in a usual manner, it is just allowed to ferment (about 30 mins) for a brief period to recover from the strains of mixing.

Since the dough is not fermented, the two functions of fermentation (i.e. production of gas and conditioning of gluten) are achieved by increasing the yeast content (2 to 3 times) and by making the dough little warmer and slacker.

Even though it is possible to prepare the bread which is fairly acceptable especially during emergency situations, due to the absence of fermentation the gluten and starch are not conditioned well to retain the moisture.

4. Delayed Salt Method:-

This is a variation of straight dough method, where all ingredients are mixed except salt and fat. As salt has a controlling effect on enzyme action of yeast, the speed of fermentation of a salt less dough will be faster and reduction in total time of fermentation will be faster.

Salt is added at a knock back stage. Method of adding salt at later stage may be according to the convenience of individual baker. It may be sifted on the dough and mixed or may be creamed with fat and salt.

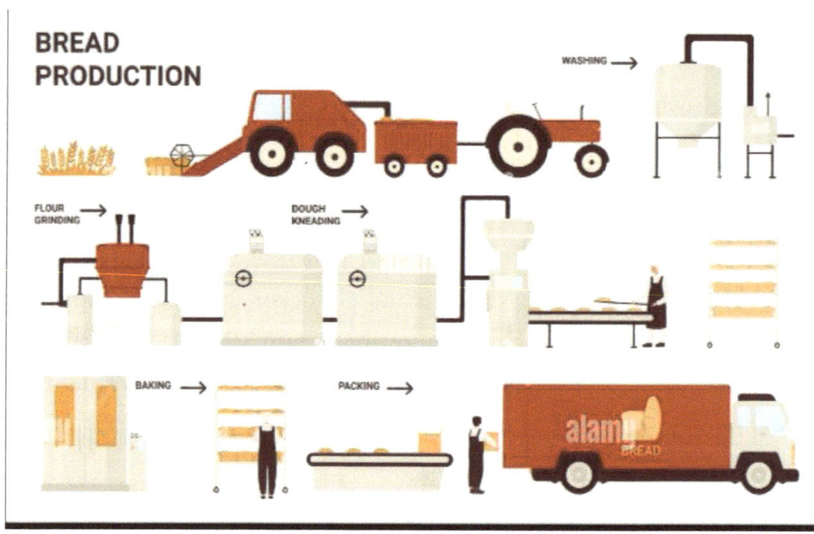

A factory layout of Bread Processing Plant

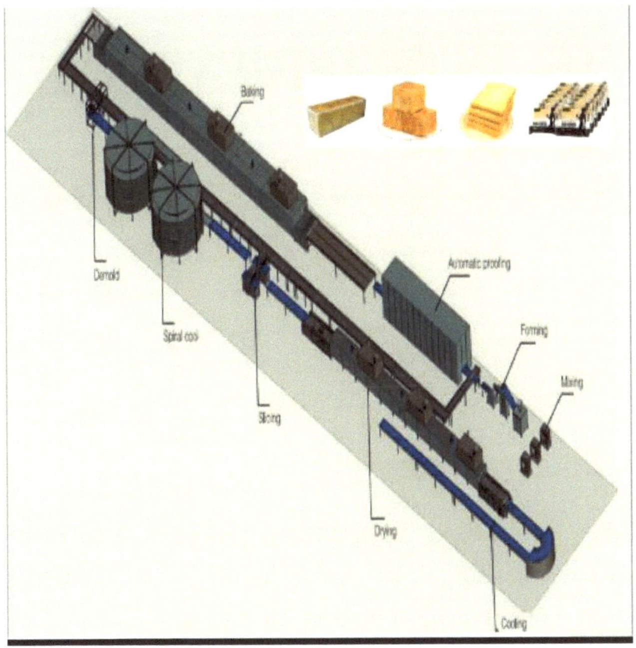

Automated Toast Rusk Production Line

CHAPTER 3

Food Additives and Bread Keeping Quality

3.1 Food Additives in Breadmaking

1. Emulsifiers:- Emulsifiers are common additives used in breadmaking and are classified according to two main functions: (i) crumb softeners and (ii) dough conditioners or gluten strengtheners.

Mono and di glycerides are main examples of crumb softeners while diacetyl tartaric acid (DATA) esters of mono and diglycerides (DATEM) and polysorbate are two prominent examples of dough conditioners.

Mono and *di-glycerides* and their derivatives account for about 70% of food emulsifiers production. Bakery is by far the field of greatest application. Mono and diglycerides are generally manufactured by esterification of triglycerides with glycerol, yielding a mixture of mono, di and triglycerides. The hardness of monoglyceride is mainly determined by hardness of edible fat from which monoglyceride has been produced. Since monoglycerides are functional part, molecular distillation can be carried out to increase their concentration.

Distilled monoglycerides are used as anti-staling agents in bread, as they soften crumb of the product after baking and retain this softness during beginning of shelf life. They act by binding to the amylose fraction of wheat starch at high temperatures typical of baking. In doing so, they slow down retrogradation of starch during cooling and storage. Distilled monoglycerides have the greatest effect on softness compared to other types of emulsifiers and less effect on loaf volume. Result is a fine crumb with considerable elasticity. Optimal dosage is 0.2%.

Lactylates are another group of emulsifiers where the fatty acid represents the non-polar portion and ionic lactic acid polymer represents polar portion. Calcium stearoyl-lactylate (CSL) and sodium-stearoyl lactylate (SSL) are typical dough conditioners. Both are commonly used in manufacturing of white bread and are employed as dough strengtheners. They act as anti-staling agents, aeration aids and starch/protein complexing agents. They carry a high degree of hydrophilicity and their salts hydrate readily in water at room temperature. The sodium salts hydrate more rapidly than calcium salts, giving SSL and CSL different functionalities in short baking processes. The strengthening effect of lactylates relates to their ability to aggregate proteins which helps in the formation of gluten matrix. They interact with proteins through: i) hydrophobic bonds between the non-polar regions of proteins and stearic acid moiety of lactylates and ii) ionic interactions between the charged amino acid residues of proteins and carboxylic portion of lactylates. In case of bread doughs, these effects result in increased dough viscosity, better gas retention and ultimately greater bread volume. *Effect of lactylates on dough handling properties and proofed dough volume are related to protein complexing.* As proofed dough is heated in early baking phase, lactylates are transferred from protein to starch. Coating on starch significantly delays starch gelatinization, keeping the viscosity low and allowing additional expansion of dough in oven. Resultant dough is thus softer than unemulsified dough and allows more abusive mechanical working without causing irreversible damage to the protein structure. *Both CSL and SSL provide good yeast raised dough strengthening effect.*

Polysorbates are sorbitol derivatives and form part of a group of emulsifiers known as sorbitan esters which can be further modified to polysorbates. These are added as dough strengtheners to improve baking performance. They stabilize dough during late proofing and early stages of baking where there are great stresses on inflating cells. Their use results in loaves with greater volume and a fine and uniform crumb structure.

2. Hydrocolloids:- Hydrocolloids act as stabilizers, thickeners and gelling agents, affecting the stabilization of emulsions, suspensions, and foams and modifying starch gelatinization. Crumb structure and texture are positively influenced by presence of gums by controlling the granule swelling with the help of hydrocolloids which interacts with molecules leached out from starch granules. Hydrocolloids are very important as breadmaking improvers as they enhance dough handling properties, improve quality of freesh bread and extend shelf life of stored bread.

3. Enzymes:- Fungal amylases and beta-amylases are used as enzymes in breadmaking. Glucoamylase and bacterial maltogenic alpha-amylase are also used.

4. Preservatives:- Propionates, Sorbates and Acetates are used as bread preservatives.

3.2 Other Breads

Brown Bread

These are wholemeal breads in which the level of fat is raised to about 1.5% as compared to 1% in white bread. A short fermentation is used for wholemeal bread where the dough is allowed to ferment for 1 hour at an appropriate yeast level and temperature.

Whole Wheat Bread

Whole wheat bread is made using flour which includes entire wheat kernel, including the bran, germ and endosperm. Whole wheat bread is known for its heartier texture and a richer, nuttier flavor compared to white bread. It is prized for its nutritional content due to the inclusion of bran and germ.

Gluten Bread

Breads fortified with vital gluten, other protein sources like whey extract, casein, yeast, soya flour.

High Fibre Bread

This bread has both high fibre content and fewer calories per unit . The high fibre content is achieved by addition of various supplements such as kibbled wheat, wheat bran, resistant starch, powdered cellulose.

Wheat bran comprises of approximately 12-15% of the wheat grain. Consumption of bran offers many health benefits such as reduced risk of certain types of cancer, promoting positive health effects on GI tract, decreasing intestinal transit time, increasing fecal bulk and stool number, preventing and treating constipation, treating diverticulosis and IBS (Irritable Bowel Syndrome), reducing obesity risk and weight management, protection against gallstone formation, affording significant benefits to diabetics by improving glycemic control and reducing requirements for insulin.

Portion of starch and starch products which resist digestion in the small intestine are described as Resistant Starch (RS). There are different categories of Resistant Starch: RS1(physically inaccessible), RS2 (compact granular structure), RS3 (retrograded or crystalline non-granular), RS4 (chemically-modified or re- polymerized) or amylose- lipid complexed (RS-5) starches. RS may be categorized as functional dietary fibre, and can entirely pass the small intestine and can behave as a substrate for growth of probiotic microorganisms. RS2, a high amylose starch is a prebiotic.

3.3 Bread Quality Measurement

Farinogram is a graphic representation of a dough consistency (C) as a function of mixing time (t). The C values, expressed in farinograph units (FU), are proportional to magnitude of the torque of stirrer arm, which balances resistance of dough to shear stress. A typical farinograph is divided into two parts, a dough development stage characterized by an increase in consistency C(t), and second part exhibits decline in consistency C(t) caused by overmixing and is called dough softening stage. The Dough development stage is influenced by two basic processes namely hydration i.e. water absorption by the flour components and a structure-formation process. The latter involves initiation of intermolecular bonds between gluten proteins, mainly disulphide, and gradual formation of a spatially continuous gluten network binding starch and other flour components.

Detrimental effects on dough handling and bread quality due to flour replacement with dietary fiber

The addition of fiber enhances the nutritional quality of bread but could alter the technological quality and sensory characteristics of the final product as well as interfere with the processing. Response Surface Methodology (RSM) is used to study effect of different dietary fibre sources on farinographic parameters of wheat flour.

Water absorption and mixing properties of the wheat flour are determined in a Brabender Farinograph according to Approved Method 54-21.01 (AACC International 2010) using 300 gms of flour. Farinographic parameters obtained from farinograms are water absorption (WA), arrival time (AT), dough development time (DDT), departure time (DT), dough stability (DS), and mixing tolerance index (MTI).

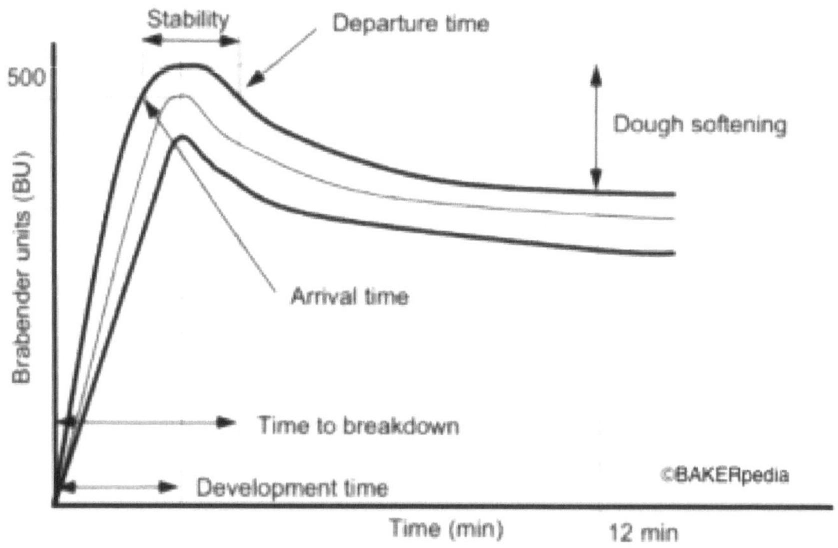

A typical representation of a Farinograph

Extensograph is a device for testing the extensibility properties of wheat dough. With the help of the dough extensograph testing we can understand three critical parameters: resistance, extensibility and energy. Resistance measures the amount of force required to stretch the dough, offering insights into the dough strength and elasticity. Extensibility indicates dough's capacity to stretch which is vital for predicting its performance during baking, impacting loaf volume and crumb texture. An extensograph device consists of several components, including dough holder, stretching arm, and measurement devices which record the dough's response to applied stress. To conduct a test, prepared dough sample is placed in the dough holder, and the stretching arm gently pulls the dough until it breaks. Preparing the dough requires specific mixing and resting protocols to standardize moisture content and gluten development.

High resistance indicates a stronger dough, suitable for products requiring significant structure such as certain breads. High extensibility signifies more stretchable dough, ideal for items requiring stretchability such as pizzas or pastries. A balanced ratio between resistance and extensibility points towards a well rounded dough suitable for multiple applications.

An *alveograph* measures resistance of dough to extension and extent to which it can be stretched under the method conditions. A sheet of dough of definite thickness prepared under specified conditions is expanded by air pressure into a bubble until it is ruptured. The internal pressure in bubble is graphically recorded on moving paper.

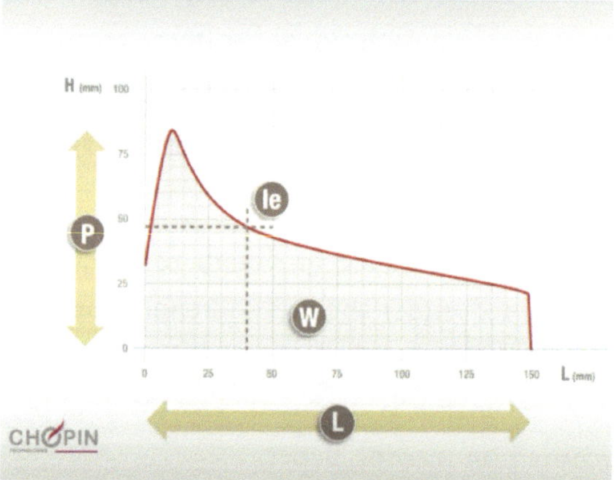

Characteristic curve of an Alveograph

P represents Tenacity of the Dough (Capacity to resist deformation)

L represents Extensibility of the Dough (Maximum volume of air the bubble can contain)

P/L: Curve configuration ratio

le represents elasticity index

W represents baking strength(area under curve) or Energy value

Baking Strength, W depends upon several factors such as the Protein Quality, Starch damage, Enzymes, Interactions. W is used widely in alveograph for predicting processing behaviour of flours.

A slower hydration implies longer duration of the formation of gluten network.

However, the health promoting additives can dramatically alter the rheological behaviour of bread dough by influencing hydration and conformational changes in gluten proteins. Dietary fibers are characterized by increased water absorption, the greater degree of water absorption in comparison with gluten proteins can induce a dehydration phenomenon during mixing. There could be migration of water from flour components, primarily gluten and starch, to fibre particles leading to partial dehydration. A poorly hydrated gluten network becomes less elastic and more resistant to mixing. Fibre substantially modifies the conformational structure of proteins during dough development, there is increase in the proportion of hydrogen bonded β-sheets due to aggregation or abnormal folding. Structural modifications lead to an impact on the viscoelasticity of the gluten network, making it stiffer.

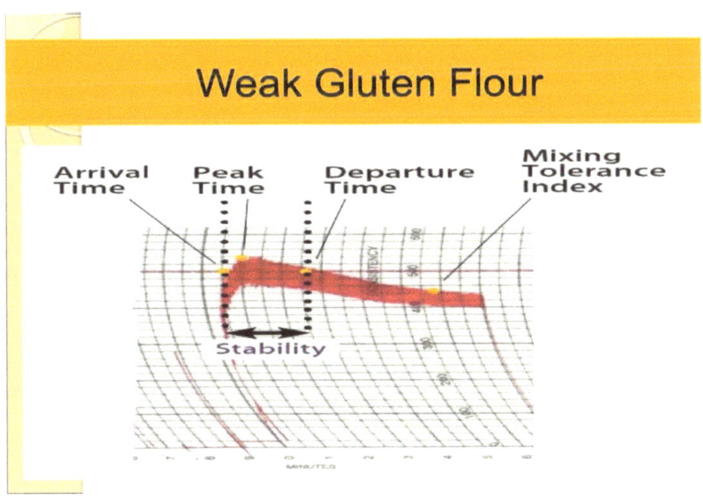

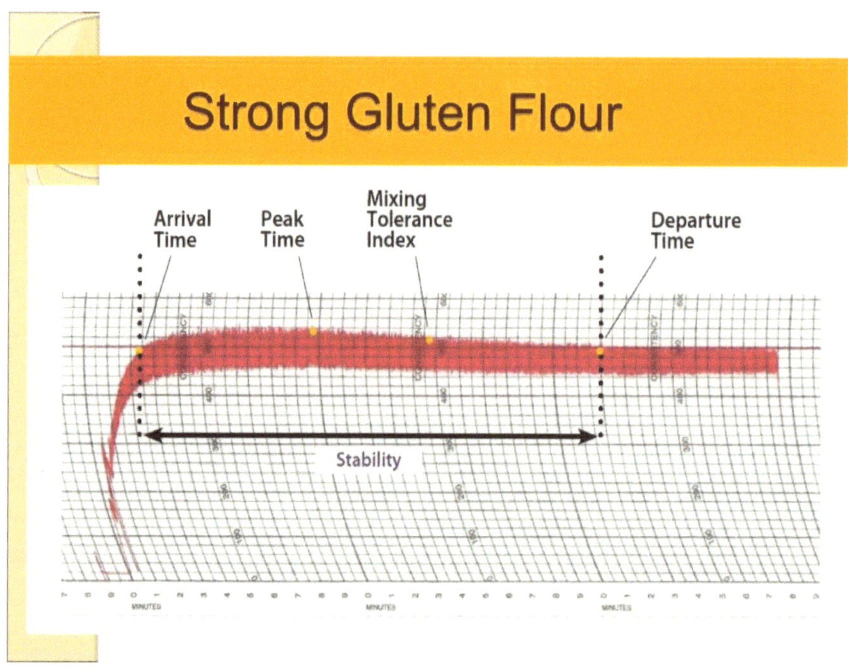

Granary Bread

Propietary bread made from mixture of wheat and rye which is allowed to sprout, kiln dried and rolled. Barley malt is added to it.

Rye Bread

Rye Bread is made with various proportions of flour from rye grain. It can be light or dark in color depending on the type of flour used. The flavour of rye bread is distinctive and texture denser as compared to wheat bread. Rye bread is typically made using a sourdough starter, which provides a tangy flavor and and contributes to its leavening. The dough is mixed with water, salt, and sometimes a small amount of wheat flour. After mixing, it is allowed to ferment, often for an extended period which could range from several hours to several days. Dough is then shaped into loaves and baked. Rye bread comes in various forms from dark to hearty to lighter versions, and it is known for its earthy, slightly sour taste.

Darker rye bread is a great complement to hearty soups and stews, while lighter rye bread can be enjoyed with spreads like butter, cream cheese, or smoked salmon.

3.4 Wheat Types

Wheat is categorized as hard or soft based on the milling characteristics, related to the way the endosperm breaks down. In hard wheats, fragmentation of the endosperm tends to occur along lines of cell boundaries while endosperm of soft wheat fractures in a random way. 'Hardness' is related to the degree of adhesion between starch granules and surrounding protein, differences in endosperm texture must be related to differences in the nature of the interface between starch and protein matrix.

Three species of wheat namely T. *aestivum*, T. *durum* and T.*dicoccum* are cultivated in the country. Quality requirements of wheat for various products like chapati, bread, biscuit, pasta are different.

For bread making, its best to use a "hard" variety of wheat, such as hard red wheat or hard white wheat. Hard wheat is high in gluten, a protein that becomes stretchy on kneading. This stretchy gluten captures tiny gas bubbles which yeast produces within dough, which makes yeast and sourdough breads rise.

Soft white wheat is ideal for baked goods which are not kneaded like cookies and pancakes, pie crusts and crackers. Soft wheat has very low gluten content, which when used in baked goods that are not kneaded, results in tender products. It is ideal for baked goods which have a tender crumb like muffins, pancakes, quick breads, biscuits and pie crusts.

3.5 Regulations on Bread

Labelling Requirements under FSSAI

The nomenclature of various breads shall comply with requirements mentioned in columns (3) and (4).

S.No. (1)	Name of Bread (2)	Specialty Ingredient (3)	Minimum Amount of Specialty Ingredient as % of Flour (4)
1.	Whole Wheat Bread	Whole wheat Flour (Atta)	Whole wheat flour should be at least 75%
	Wheat Bread or Brown Bread	Whole wheat Flour (Atta)	Whole Wheat Flour should be at least 50%
3.	White Bread	Refined Wheat Flour (Maida)	-
4.	Multigrain Bread	Food Grains permitted under FSS (Food Product Standards & Food Additives) Regulations, 2011	Minimum 20% should be grains other than wheat
5.	Specialty Bread	Specialty ingredient that must be present in case a prefix is added to the term "bread" on the label	
(a)	Oatmeal Bread	Flours used must have at least 15 percent oats content on a 100 gms basis	
(b)	Milk Bread	Milk Solids	6%
(c)	Honey Bread	Honey	5%
(d)	Cheese Bread	Cheese	10%

S.No. (1)	Name of Bread (2)	Specialty Ingredient (3)	Minimum Amount of Specialty Ingredient as % of Flour (4)
(e)	Rye Bread	Rye Flour	20%
(f)	Bran Bread	Edible Bran	5%
(g)	Protein enriched bread (Protein prachur bread)	Edible Protein	15% Protein
(h)	Bread such as garlic bread, masala bread, oregano bread etc.	Garlic, Oregano etc.	2%

Calculation of specialty ingredient is on 100g flour basis

3.6 Standard of Bread

Sold as white bread or wheat bread or fruity bread or bun or masala bread or milk bread, means product is prepared from a mixture of wheat atta, maida, water, salt, yeast or other fermented medium containing one or more of the following ingredients:

Condensed milk, milk powder (whole or skimmed), whey, curd, gluten, sugar, Gur or Jaggery, khandsari, honey, liquid glucose, malt products, edible starches and flour, edible groundnut flour, edible soya flour, protein concentrates and isolates, vanaspati, margarine, refined edible oil of suitable type or butter or ghee or their mixture, albumin, lime water, lysine, vitamins, spices, condiments or their extracts, fruit and fruit product (Candied and crystallized or glazed), nuts, nut products, oligofructose (max 15%) and vinegar.

3.7 Bread Spoilage

Bread is subjected to a number of changes during storage once it leaves the oven. A number of changes leading to the loss of organoleptic freshness are identified in bread storage. 2 types of factors are responsible for freshness loss in bread: attributable to microbial attack & attributable to chemical or physical changes leading to progressive firming up of crumb, referred to as "Staling."

Microbial Spoilage of Bread:-

The most common forms of bread deterioration are *moisture loss* and *microbiological spoilage.*

Most common microbiological spoilage problem in bread is mould growth. A little less common problem is the bacterial spoilage condition known as 'rope' caused by Bacillus species. Yeast spoilage is the least common microbial spoilage in bread.

Mould spoilage is due to the processing conditions in bread manufacturing. As the bread is baked, bread loaves are free of mould or mould spores due to thermal inactivation in the baking process. However, once the bread is exposed to the processing environment, it becomes contaminated from the mould spores present in the atmosphere during the processes of cooling, slicing, packaging and storage.

Due to the non-sterile environment inside the processing facility, mold spores spread easily in the environment through dry ingredients especially flour, flour dust which tends to spread easily. One gm of flour contains around 8000 mould spores, so one can imagine the susceptibility of the bread towards mould growth. In large bakeries, this exposure is made to control by segregating the flour handling areas from the cooling and packaging areas of finished bread. Moulds tend to grow in humid atmosphere, especially on a loaf inside a wrapper. This type of growth is more likely to occur if the bread is wrapped hot from the oven so that droplets of water condense on the inside surface of the wrapper. Once the bread is cut, more susceptible surfaces are exposed to mould infection. Hence, a sliced wrapped bread, due to the moist cut surface are more susceptible towards spoilage but a proper packaging prevents moisture loss.

Common bread spoilage moulds include *Penicillium* spp., wheat breads spoilage moulds include *Penicillium, Aspergillus, Cladosporium, Mucorales, Neurospora.*

Some moulds present a severe risk to public health because they produce mycotoxins. Exposure to mycotoxins occurs either by directly eating bread spoiled by mycotoxigenic moulds or indirectly as a result of people consuming products of animals fed contaminated bread.

Methods for limiting microbial spoilage of bread are reviewed by some authors. These are associated with firming of the bread crumb through moisture loss but could occur in the absence of moisture loss through the movement of water between bread components at the microscopic level.

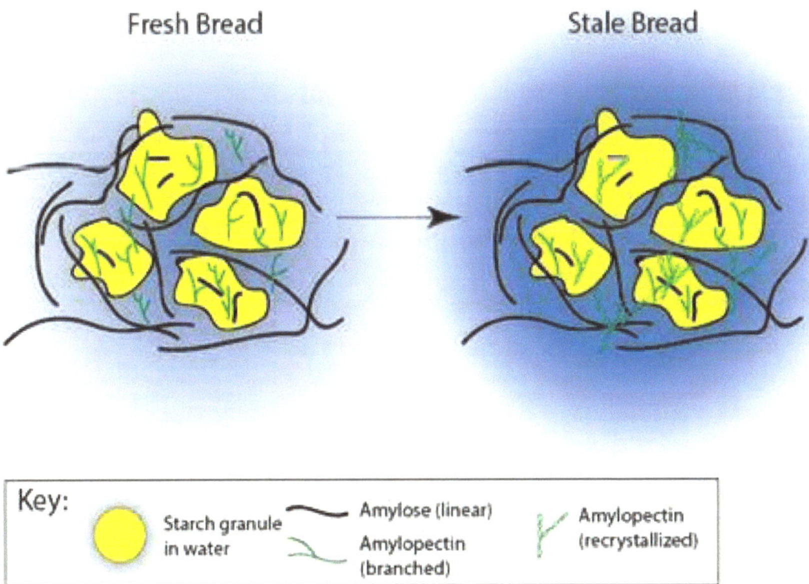

Some noteworthy points regarding mould growth in breads:-

1. Brown and wholemeal breads appear to become mouldy rather than white breads because mould growth is more visible on darker surfaces.

2. The Processing method also has an effect on the rate of mould growth. For example, breads made with Chorleywood Breadmaking Process (CBP) and Activated Dough Development (ADD) have a slightly shorter shelf life than bulk fermented bread.

3. *Rhizopus stolonifer* is a common black bread mould. *Neurospora Sitophila* is another type of mold reddish in colour, and found in bread stored at high humidity. 90% of mould isolated from a range of bread in Northern Island were *Penicillium spp*.

Bacterial Spoilage

Rope is a spoilage problem of bread and occurs in bakery products with high equilibrium relative humidity (ERH), which is greater than 90%. It is caused by mucoid variant of *Bacillus subtillis* or *Bacillus licheniformis.*

Primary contamination is from raw ingredients which may be present in or on the equipment. *B. subtilis* appears naturally in soil and hence rope bacteria may be present on the outer parts of grains and vegetables.

Preservatives in control of bread spoilage

Mold Spoilage is a serious problem, and use of preservatives is an attractive means to diminish spoilage and ensure food safety.

There are different kinds of preservatives, both Natural and Chemical. The main work of preservatives include: Strengthening of the overall integrity of the loaf, slowing down mold growth, and fighting off bacteria.

Chemical additions to bread are ongoing for a long time but the practice is debated. Many research organisations question the safety of these additives and have lobbied for more "Natural" selections. Food regulations have also placed certain limits on the use of preservatives and which all contents are permitted.

Reduction of preservatives to sub-inhibitory levels has been shown to stimulate growth of spoilage fungi in some cases and stimulate mycotoxin production.

Spores survive baking, and germinate and grow within 36-48 hrs inside loaf to form characteristic soft, stingy, brown mass with odour of pineapple or melon. Volatile compounds are released including diacetil, acetoin, acetaldehyde and isovaler-aldehyde which contributes towards the typical sweetish odour of rope.

3.9 Packaging of Bread

Regular bread requires techniques and materials which preserve loaf volume.

Baked Good	Type of Packaging	Benefits	Drawbacks
Bread	Plastic bag, typically LDPE bag sealed with twisted tag	Retards moisture loss	Moisture migration from crumb to crust generating a leathery crust
Specialty Bread	Plastic bag, typically an OPP or PET bag with perforations.	Retains crisp crust	May pose contamination risk due to film perforations

The food revolution in packaging aids in development of packaging techniques such as active packaging (AP), and intelligent packaging (IP). There is effective interaction between packaging material and antimicrobial component forming a protection layer around the food material and its surroundings which increases shelf-life, decreases sensory losses and enhance the hygiene, safety and security of food product. Modified Atmospheric Packaging (MAP), works on replacing air with a mixture of carbon dioxide (CO_2), nitrogen and inert gas in packaging mainly decreasing the O_2 content to less than 1% of the surface of packed food. Fungistatic property of CO_2 inhibit the growth of mold by changing the metabolic activity of cell membrane, resulting in detrimental effect on growth. However, there are varying conflicts regarding the use of MAP packaging, stating that high concentration of CO_2 increases the acidity of baked food affecting taste and sensory properties.

EFSA has developed new and innovative concepts for packing foods such as AP and IP, permitting interaction between food and packing material by carrying functional material and deliberately interact with food substances and protect food from external contamination.

Table 1 mentions the main functions of Active and Intelligent Packaging

Components of Active Packaging	**Components of Intelligent/Smart Packaging**
Consists of various chemical components which have huge anti-fungal/microbial properties, incorporated in the form of sachets/ films.	Packaging material capable of indicating time temperature for identifying contamination.
Incorporating ethylene scavengers/ ethanol emitters/oxygen absorbers in packaging sachets/films.	Food Products with sensors/ indicators in controlling micro spoilage
Packing contains inbuilt heating or cooling mechanism according to functions and requirements	Shock indicators for identifying physical damage to food products
Moisture absorbing/carbon dioxide emitters in the form of pads and sheets in food products	Food Packaging sensor which distinguishes between allergen and non-allergen
Sachets/films made to remove/ absorb odor and flavor.	Leakage indicators with microbial growth sensors
Contaminants causing food spoilage organisms are controlled in the form of sachets, films, and coating.	Identification of indicators/sensors for contaminants

3.10 Research Trends in Bread

Functional Breads:-

The production of functional breads has become a trend. Flours with high amylose content have gained interest due to their functional properties. High amylose starch can retain its crystallinity which resists to amylase degradation contributing towards slow digestibility. Starch escaping digestion in small intestine is recognized as Resistant Starch (RS).

RS provides a carbon source to colonic microorganisms and is fermented to short chain fatty acids which are known to have several health benefits, it is also considered as a dietary fiber.

Thus, the introduction of Resistant Starch for breadmaking is a method for increasing the fiber content of bread. Increased amylose content is related with slower digestion and greater RS, so high amylose starch helps in improved nutritional properties of bread.

There is development of assortment of functional bakery products of increased nutritional and biological value for dietary purpose with the use of natural food fortifiers (based on sprouted dispersed grains of rye or wheat), characterized by presence of vitamins, minerals in a biodigestible form, essential amino acids etc.

Manufacturers are launching fortified breads and are incorporating more wholesome flour blends. Some of the most desired components are lentils and chickpea flour due to their high protein and fiber content. Protein breads have recently made a niche in the market with several companies launching interesting protein blends. For a good claim, the bread must contain 10% to 19% of the Daily Value of Protein. For an excellent claim, it's 20% or more. The percent Daily value of Proteins is determined using PDCAAS. This translates to 5 grams or 10 grams of complete protein per serving for each respective claim. When implying a protein claim, the US FDA required the inclusion of the percent Daily value. Now, let us understand the relationship between complete proteins and percent Daily value. A bread containing 10 grams of complete protein may make an excellent source of protein claim. A bread with 10 grams of protein from peas and nuts – incomplete protein- most likely would qualify for a "good source of protein" claim, and further should not flag 10 grams of protein per serving as it is misleading. Animal and Quinoa are complete proteins. Plant proteins are incomplete hence bakers opt for blending them in their formulations.

Supplemental nutrients like vitamins, biotics, and useful substances like caffeine are also making their way to the market.

Use of organic preservatives has the feasibility of satisfying criterias including productivity of goods and meeting wholesomeness for consumers. Chemical preservatives such as propionates and sorbates in bread and other bakery products is of considerable interest.

Chemical preservatives such as calcium propionate are commonly applied to expand the microbial lifespan of bread. Using bio-preservation techniques on bread can help in solving the issue of health risks posed by chemical preservatives. Bio- Preservatives are organic resources which can be utilized to eliminate microbial populations while improving food quality. It is a revolutionary concept for preserving fresh goods and is recognized as GRAS.

Bio-Preservatives have emerged as a solution to generate "clean-label" foods. Bio- Preservatives can be adopted as anti-fungal substances to prevent fungal degradation and prolong shelf-life, reducing public health risks. Bio-Preservatives such as lactic acid bacteria, essential oils or natural nanoparticles are becoming popular. A good biopreservative has the following characteristics:-

Has expansive antibacterial spectral range, is non-toxic to humans, suitable for lower doses, have slight impact on product pH, not impair product odor, colour, flavor, have high water solubility, non-corrosive, be unreactive, and have no detrimental effects on fermentation.

Some Recent Research News Articles

Waste Utilization:- A research published in the Journal Frontiers of Microbiology suggests use of spoiled bread as a medium for cultivating microbial starters for the food industry.

Bread waste creates both economic and environmental impact as most waste ends up in landfill which emit greenhouse gas like carbon dioxide and methane. The idea behind the research is to repurpose all the discarded dough to feed the microorganisms needed to kickstart fermentation in food industries like bakery, dairy and wine industries. As of *2020*, this process of combining the need for disposing off the bread waste with that of cheap source for media production, with fitting for cultivation of several food industry starters, is patent pending.

Bread waste is comprised of 50-70% starch which is a homopolysaccharide consisting of two components amylose and amylopectin. In order to unravel the potential of BW as sustainable feedstock, it should be subjected to pretreatment or hydrolysis for the breakdown of starch into monomeric fermentable sugars.

Bread waste is comprised of 50-70% starch which is a homopolysaccharide consisting of two components amylose and amylopectin. In order to unravel the potential of BW as sustainable feedstock, it should be subjected to pretreatment or hydrolysis for the breakdown of starch into monomeric fermentable sugars.

Use of BW as a feedstock for biorefineries has multiple advantages including financial and economical benefits to industries generating BW. The low cost of BW will deliver financial benefits to biorefinery plants through decreased cost associated with use of waste resource as microbial feedstock.

Novel Preservation technologies for preservation of Bread:-

Consumer interests and demands have forced bakeries to invest in preservative free technologies such as vacuum cooling, infrared, cold-plasma and ultraviolet light. A study involving application of UV-C radiation at 32 kGy confirmed the potential of increasing shelf life of product by 100% compared to 26% using calcium propionate.

New improved technologies have introduced installation of spiral coolers operating in continuous mode to provide seamless transportation experience of baked goods from oven to the slicing and packaging areas thus reducing the risk of mould growth.

3.12 Bakery Industry in India

Bakery industry in India holds a significant position in the food processing sector. As per an Expert Market Research website, Bakery industry in India is valued at approximately USD 11.07 billion in 2024. It is expected to grow and reach at USD 27.43 billion USD in 2034.

Bread and biscuits account for over 82 percent of total bakery products produced in the country. India is in fact the third largest biscuit manufacturing company after USA and China. Busy lifestyles, expansion of fast food chains, changing eating habits and Western influence are factors contributing towards the growth of this industry.

Expansion of distribution channels such as supermarkets and online stores are driving the demand for packaged bakery products. Key trends in the India bakery industry are growing health concerns, innovations in bakery products, and development of sugar free, gluten free and fortified products.

Challenges impacting this sector include maintaining shelf life of products, particularly in remote areas. The other challenge is maintaining consistent product quality from unorganized players.

Bakery Business is a very profitable industry in India with many startups already operational and launching innovative products.

References

1. TECHNOLOGY OF BREAD MAKING: e-Krishi Shiksha
2. Technology of Breadmaking- Stanley P. Cauvain
3. Bread Baking- Daniel T. DiMuzio
4. Food Additives and Processing Aids used in Breadmaking | IntechOpen
5. BakersPedia

www.ingramcontent.com/pod-product-compliance
Ingram Content Group UK Ltd.
Pitfield, Milton Keynes, MK11 3LW, UK
UKHW060405300726
14090UKWH00006B/453

* 9 7 9 8 8 9 9 2 9 9 6 4 3 *